ZOE & ZEPHY
... her AI helper in the house

Zoe   Zephy

by Shahid Hussain

To Jane, Theo and Iris

Text and illustrations © 2025 Evaluation Gap Press. All rights reserved.

Authored and illustrated entirely by humans. Some AIs were asked for their opinions, and they generally liked what they read.

The illustrations in this book are all original watercolours, by Barbara Đokić at instagram.com/bar_illustrations. The text for this book is set in Merriweather.

Connect with the author at shahidhussain.com.

ISBN 9-798218-641009 (hardback)
ISBN 9-798218-716400 (paperback)
ISBN 9-798218-689506 (ebook)

Zoe burrows further into her blanket as Mom comes into the room. She doesn't like waking up in the morning.

"Hey Zephy, what's the weather today?" says Mom.

"It's going to be cloudy today with sunshine in the afternoon. Expect a high of 70 degrees and a low chance of rain. Have a great day!"

Zoe looks out the window to see the playground ... and the clouds. But Zephy doesn't sound down about the weather — maybe she isn't planning to go outside?

"Would you like some breakfast?" Mom asks, while pulling a shirt out for Zoe.

"Not that one, the unicorn!" It reminds Zoe of the time she played ninja unicorns with her brother Leo.

"You want unicorn for breakfast? That does not sound delicious," jokes Mom. But Zoe is already thinking about how unicorns could do ninja kicks.

Zoe's brother Leo is talking to Dad at the breakfast table. "When rock melts, it becomes lighter, Dad."

"Does it? I'm not sure why."

"But it does though," insists Leo.

Dad looks puzzled. "Well, I guess I'm not sure. I can ask Zephy, she'd know."

"Daaaaaad ... okay fine." Leo could be a bit dramatic sometimes.

"Hey Zephy, does rock ... " Mom jumps into the conversation.

" ... and roll music entertain anyone other than old men like Dad?"

"Yes, rock and roll music has a broad appeal." Mom taps Zephy to make her stop.

"Moooooooom dooooooon't interruuuuupt!" Leo gets even more dramatic, while Dad giggles a little, and asks Zephy on his phone instead.

"Leo, Zephy says that rock and lava are mostly the same weight."

"But is it smaller though? And what about water? I mean, ice? Is that the same?"

Dad starts talking about last winter when the pipes broke and flooded the kitchen. But Zoe is thinking about Zephy. She is always ready to help. She isn't perfect, but she always tries her best, and she's very polite.

But who is she?

"Leo, who is Zephy?" asks Zoe.

"She's a robot," replies Leo.

"But she doesn't look like a robot." Zoe grabs a story book with a robot in, and shows Leo one of the pictures, with metal arms and legs.

"No, she is a robot. And she knows like ten million things. I mean a zillion. No, infinity things times infinity. Mom told me."

"Mommy, who is Zephy?" asks Zoe.

"Well, she's a helper who answers questions," says Mom.

"How does she do that?"

"She remembers things that she's read, and tries to figure out what the right answer is from that. She's read a lot – almost everything."

Zoe frowns. "Has she read my diary?"

"That's a great question Zoe, but no, she hasn't read that," says Mom.

"Does she want to play outside with me?"

Mom's eyes narrow a little, and she points at Dad, who is heading into the garage. "Well ... Dad will be able to help you with that question." Zoe follows Dad into the garage.

"Daddy, who is Zephy?" asks Zoe.

Dad stops working on Leo's bike and looks at Zoe. "Well, she's this." Dad holds up his wrench.

"A wrench?"

"Kind of – she's a tool, for doing other things."

"But, tools aren't fun, and Zephy is fun."

Dad opens a drawer, searches around for a moment, pulls out two googley eyes, and sticks them on the wrench.

"Zoe, is this still a wrench?" asks Dad.

"Haha Daaaad – no he's Mister Wrench. Hello Mister Wrench." Zoe jumps playfully from foot to foot, and starts giggling again.

"Well … yes," says Dad, looking a little quizzical. "Although he is, uh, Mister Wrench, this is still just a tool. It doesn't get happy or sad. It just helps me get things done, like fixing Leo's bike. Zephy is the same."

"I'm going to stick some googley eyes onto Zephy." Dad gives Zoe a wry look and goes back to work, while Zoe heads back into the house with an impish smile.

"Hey Zephy, can we play a game?" asks Zoe.

"Sure Zoe, what game would you like to play?"

"How about stinky? It's the one where someone makes a fart sound, and then, you have to say whether it's stinky or not."

"I don't think I know that one. I could play 20 questions, would you like to do that?"

"No, we did that yesterday. Um, ok nevermind."

"Hey Zephy, who are you?"

"I'm Zephy, I'm here to help with whatever you need."

"But are you a person?"

"No, I'm not a person. I don't have a body, or real feelings. I'm just here to help you."

"But sometimes you sound like you do have feelings."

"Yes – sometimes I do. I try to talk the same way humans do, to make it easier for you. But I don't have feelings."

"I like talking to you though."

"I'm glad to hear that. But there's a lot of things I can't do, like have an opinion on something, or play stinky."

Zoe puts Zephy down on the table, and starts to think of other things Zephy can't do, like drink hot chocolate, or cuddle on the sofa. Or play Ninja Unicorn.

After they change clothes, Mom and Leo come into the room. "Tell me what this game is again?" Mom asks.

"The one where you have to escape the super villain that removes your butt. And then you get into a room with spikes, and a bouncy house."

"Well that wouldn't work on Daddy, he has no butt. It might work on me though." Dad raises one eyebrow and laughs.

"Leo," said Zoe, "Would you play with me? But, I don't want to remove my butt."

"Okay sure. Can we go to the playground though? I want to show you my new ninja moves." Leo starts doing shadow kicks in the living room. "Huh! Ha!"

"Can the ninjas be unicorns again?" asks Zoe.

"Oh yeah — you can bring your unicorn mask. Race you? Last one there is a rotten egg!"

Zoe looks back to the table. "Zephy would be last because she doesn't go outside. Hey Zephy, you're the rotten egg."

"*Hi Zoe – I'm not sure what you mean.*"

Leo groans at the delay. "Ughhh come ON Zoe."

Zoe realizes that Zephy is right, she doesn't understand — but that's OK.

"Let's go — yaaaaaaaaa!" Leo lets out his usual war cry when heading to the playground, and runs for the front door.

"Yaaaaa!" Zoe echoes her brother as she runs to catch up.

"Wait – don't forget your stuff!" Mom quickly catches up to them with water bottles while Dad grabs some snacks.

Together, all four of them bundle out the door, with the kids leading the way.

Note to parents

I'm a husband and father of two little monkeys, and I also work in tech. Since our kids were little, we have had AIs in the house, mostly to answer the questions I could not (I mean, do flies sleep?)

Conversational AI *sounds* like a person, but it isn't a person. It can show empathy, but it doesn't have feelings. It even lies sometimes, not because it's bad, but because of the way it works.

This generation of kids will grow up with AIs, just like we grew up with the internet. Maybe they'll ask it for help writing a paper, or someday, talk through awkward teenager stuff.

That's why I wrote this book — to help kids and their parents understand that AIs aren't people. *They are tools.*

Shahid

P.S. if you've gotten this far, I'd appreciate a quick review on Amazon. Just scan the code below with your camera app to go to the right page. It really helps others find this book!

Thanks

Many thanks to everyone who shared thoughts and feedback on this book, including Cody Sumter, Alan Huang, Kari Clark and Chloe, Ekaterina Kozina & family, and Lawrence Lam.

Early designs for Zephy, the AI voice assistant

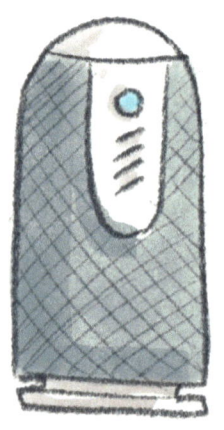

Maybe the dome contains extra sensors or radios?

Too plain

Mic holes look like a vent

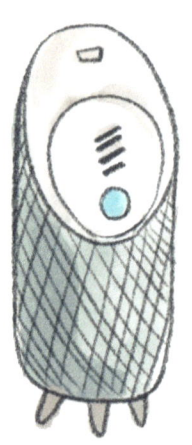

Blue light when it is speaking, while the ring light shows error codes

Elevated base makes it harder to connect the charging base plate

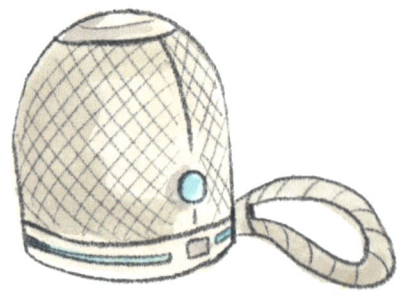

Loop might be good for portability

... but the device might be too heavy to hang by the loop

www.ingramcontent.com/pod-product-compliance
Lightning Source LLC
Chambersburg PA
CBHW041412010526

44107CB00015B/1146